Factors of Fractions, Units of Units Compared to Length, and the Equilibrium of pi Over the Period of pi.

By: Stephen John Macko, B.A.

Table of Contents

Introduction:

It is the purpose of mathematics to produce within a student
the desire to learn and to understand. The objective of this
book is to allow the student to understand how number
produces answers that are based on and equilibrium of
problems to solutions.

By: Stephen John Macko, B.A.

I.   A variable taken out of a fraction and the observation of length and factors:

The realization that the denominator of a fraction denotes the numerator, a variable is taken out of fractions (ie. 1/2, 2/4, 3/6, etc…) when the answer is multiplied times the denominator and then divided by the denominator.

Example:

½ = (.5 x 2)/2;

2/4 = (.5 x 4)/4;

3/6 = (.5 x 6)/6; etc…

1/3 = [(.33 x 3) + (1/100)]/3 This calculation proves a remainder proof.

This is very important in length. The possibility to have 2 inches divided by 4 inches compared to 1 inch divided by 2 inches proves how factors of fractions are important to mathematics, engineering, and building.

1x1)/1=(1/1)x1

(1x1)/1=(1x1)/1

(1/1)x1=(1/1)x1

Notes:

II. The integration of the inch and the meter calculated to $1/16^{th}$ and $2/16^{ths}$ (from a fraction to a whole number):

$F(x) = x^2$

Given: x = 1/16

Metric:

Integrate: $(254/160)^2 * (1/16)^2 * (1/25.4) * (16) =$
1032256/166461440 times a factor

Inch:

Integrate: $(1/256) * (1) * (1/16) = 1/4096$ times a factor

Factor: $x^2/$Inch

Metric/Inch = 25.4

Given: x = 2/16

Metric

Integrate: $(508/160)^2 * (2/16)^2 * (1/25.4) * (16) =$
16516096/166461440 times a factor

Inch:

Integrate: $(4/256) * (2) * (2/16) = 16/4096$ times a factor

Factor: $x^2/$Inch

Metric/Inch = 25.4

The above integrations show how to integrate number based on factors.

The law of ratios is the largest number divided by the largest number compared to a smaller number divided by the largest number.

Notes:

Notes:

III.   pi over the period of pi:

The relationship of units per inch and pi is as follows:

$pi_x{}^2/9.921875 =$ x units per inch/16

The proof that the area of a circle within a circle over the period of pi is proven to be equal in area.

Therefore knowing the radius or the pi of x units per inch is based on area as well:

$r^2 * pi_x = (1)^2 * 3.141592654$

This is a solution that pi is found by finding the units of pi per inch.

Therefore:

$15.7509765625/9.921875 = 25.4/16$

Notes:

Notes:

Notes:

IV.   `Notes:

$[1032256/166461440]/[1/4096] = 25.4$

Metric of x^2 where $x = 1/16$ inch:

$16*[1032256/166461440] = 0.09921875$ units^2

Inch of x^2 where $x = 1/16$ inch:

$16*[1/4096] = 1/256 = 0.00390625$ inch^2

Further notes:

$1032256/16^4 = 15.7509765625$

and

$166461440/16^6 = 9.921875$

and

$15.7509765625/9.921875 = 25.4/16$

Example of finding the units per 16 units (inch):

$\sqrt{15.7509765625} = 3.96875 = \text{pi}_{25.4 \text{ units per inch}}$

Notes:

$(1/16)^2 * 25.4 = 0.0625^2 \times 25.4 = 0.09921875$

$0.09921875 \times 256 = 25.4$ units^2

A unit that is made to equal itself when its smaller parts add to equal the entire unit, and that unit equals the unit it is converted from:

$(1/256) *256 = 1$ inch^2

Thus:

$16 \div \sqrt{25.4} = 3.1747031605302$

$3.1747031605302 \div 16 = 0.19841894753314$

$0.19841894753314^2 \times 25.4 = 1$

Then:

$25.4/(25.4/\sqrt{25.4})$

$25.4/(25.4/\sqrt[3]{25.4})$

Converted and graphed into both meters and inches, while proving a unit of comparison 1 inch = 25.4 mm

This is in comparison to converting before squaring the product.

s/r = theta/1

Notes:

$(s/s)/(r/r) = (theta/s)/(1/r)$

From 0 to 1 the second equation above allows all circles to be viewed, while the first equation refers all circles back to r = 1

Law of Ratios: The largest number divided by the largest number and a smaller number divided by the largest number.

Circles of comparison have different radiuses.

The work of the first equation is based on making 1 = 1 in conversion of area.

Area

$[(16/\sqrt{25.4})/16]^2 \times 25.4 = (1 \times 25.4)/25.4$

Volume

$\{[16/(25.4/\sqrt[3]{25.4})]/16\}^3 \times 25.4^2 = (1 \times 25.4)/25.4$

Area

Area x Length = 1

Volume

Volume x Area = 1

Notes:

$F(x) = x^2$; where $x = 16/16$

Metric:

Integrate: $(65024/2560)^2 * (16/16)^2 * (1/25.4) * (16$ units$) = (/)*($Factor$)$

Inch:

Integrate: $(256/256) * (16$ units$) * (16/16) = (/)*($Factor$)$

Interval = Factor = x^2/Inch = 1/16 units

Metric/Inch = 25.4

A double integration shows the (Factor^2)*(1/2): This is much like finding the area of a triangle 1/2ab, where a=b. This is related to the cubic relation of 25.4 units^2.

The derivative of this integration is found in units.

I have proposed that 1 inch^2 is equal to 25.4 units^2.

Please note that the integration, its' derivative, and its' double integration are all possible relations to (ie. geographic) determinations of pi.

Once again, from the above equation, 254,000,000,000 is found to be important to the metric and inch units, and, is possible to being a key relation to the number of pi itself.

Notes:

The basis:

(1x1)/1=(1/1)x1

(1x1)/1=(1x1)/1

(1/1)x1=(1/1)x1

Proves:

[F(x)*F(x)]/F(x)= [F(x)*F(x)]/F(x)

And:

[F(x)*F'(x)]/F'(x)= [F(x)*F'(x)]/F'(x)

Etc...

{[16*(1032256/166461440)]*25.4}/25.4

Notes:

Notes:

Notes:

Notes:

Notes: